In Ocean School,
one sunny day,
the fish were wondering
what they'd weigh.

They needed scales
to figure it out,
"Let's use these shells!"
said Trevor Trout.

Next they searched
the ocean bed,
gathering up
some bits of lead.

Sam Starfish said,
"Let's count to see
just how many pieces
balance me."

Next in line
was Sue Seahorse –
she didn't weigh very much,
of course.

Pete Perch jumped in,
and made a guess,
but actually
he weighed much less.

Then Bertha Bass
began to boast:
"I'm so big
I'll weigh the most."

But still to come
was Wally Whale –
he was so heavy
he broke the scale!